12th AIChE Southwest Process Technology Conference 2020

Online
1-2 October 2020

ISBN: 978-1-7138-2122-9

Printed from e-media with permission by:

Curran Associates, Inc.
57 Morehouse Lane
Red Hook, NY 12571

Some format issues inherent in the e-media version may also appear in this print version.

Printed with permission by Curran Associates, Inc. (2021)

For permission requests, please contact AIChE
at the address below.

AIChE
120 Wall Street, FL 23
New York, NY 10005-4020

Phone: (800) 242-4363
Fax: (203) 775-5177

www.aiche.org

Additional copies of this publication are available from:

Curran Associates, Inc.
57 Morehouse Lane
Red Hook, NY 12571 USA
Phone: 845-758-0400
Fax: 845-758-2633
Email: curran@proceedings.com
Web: www.proceedings.com

TABLE OF CONTENTS

CHEMICAL PROCESS TECHNOLOGY

PETROLEUM REFINING TECHNOLOGY

IIOT/BIG DATA

SUSTAINABLE TECHNOLOGY

(2a) Chemical Looping Dehydrogenation of Light Alkanes: Redox Catalyst Design and Process Analysis

Conference: Southwest Process Technology Conference

Year: 2020

Proceeding: 12th Southwest Process Technology Conference

Group: Southwest Process Technology Conference

Session: Catalysis & Reaction Engineering

Time: Thursday, October 1, 2020 - 10:30am-10:55am

Authors:

Li, F., North Carolina State University
Neal, L., North Carolina State University
Gao, Y., North Carolina State University
Haribal, V. P., North Carolina State University

Olefin productions are among the most carbon intensive processes within the chemical industry. Ethylene production by naphtha or ethane cracking is very endothermic and requires extremely high temperatures. The significant energy demand requires large amounts of fuel to be burned, resulting in up to 2 tonnes of CO_2 per tonne of ethylene produced. Similarly, production of propylene, butenes, and butadiene are also highly energy intensive. To address these challenges, we have developed chemical looping â€“ oxidative dehydrogenation (CL-ODH) approaches to convert light alkanes to alkenes. Under this scheme, an oxide based redox catalyst, also known as oxygen carrier, selectively oxidizes the hydrogen byproduct produced during dehydrogenation. When regenerated with air, the redox catalyst releases heat for an autothermal operation with high efficiency and low CO_2 emissions.

As a critical component that facilitates the entire CL-ODH process, redox catalyst development is highly important. Within this talk, the general design and optimization strategies for redox catalysts and their long-term CL-ODH performances are presented. High selectivity and olefin yields were achieved for CL-ODH of ethane, propane, and butane, making the CL-ODH system highly flexible.

To demonstrate the overall CL-ODH performance, process modeling and energy analysis results will also be discussed, showing the potential to reduce energy demand for ethylene production by up to 82%. Besides being used for centralized olefin production, a modular system that makes use of CL-ODH for liquid fuel production from distributed shale gas production sites is also briefly discussed.

(2b) Solutions for Maximum Attainable CCR Platforming Yields

Conference: Southwest Process Technology Conference

Year: 2020

Proceeding: 12th Southwest Process Technology Conference

Group: Southwest Process Technology Conference

Session: Catalysis & Reaction Engineering

Time: Thursday, October 1, 2020 - 11:00am-11:15am

Authors:

Baker, E., Honeywell UOP
Egolf, B., UOP LLC, A Honeywell Company
Trufanov, D., Honeywell UOP

Honeywell UOP has a rich history in providing customer solutions that debottleneck Platforming™ unit-specific equipment limitations. These have delivered on improved profitability from refining and petrochemical complexes, while ensuring safe and sustainable operations. However, capital utilization is a dynamic problem, with feedstocks and processing objectives responding to fluctuations in market demands.

Therefore, Honeywell UOP has introduced RMY-7 as part of a new, maximum yield performance solution. Improving the most difficult catalytic reforming reaction - paraffin dehydrocyclization - is the key. The RMY-7 solution provides customers access to a much broader range in catalyst performance characteristics, such as yield and activity, alongside unit-specific bottleneck monitoring that guides final tailoring of catalyst properties and performance. Explore in this presentation a new pathway towards dynamically enhancing reformate and hydrogen yields, where profitability and competitiveness can be enhanced within today's motor fuels market.

(2c) Modularization of Reactor Systems

Conference: Southwest Process Technology Conference

Year: 2020

Proceeding: 12th Southwest Process Technology Conference

Group: Southwest Process Technology Conference

Session: Catalysis & Reaction Engineering

Time: Thursday, October 1, 2020 - 11:30am-11:55am

Authors:

Schmidt, C., Unicat Catalyst Technologies, Inc.
Gildert, G., Vanguard Catalysts, LLC.

Modularization is a way of customizing a reaction system design, while capitalizing on economies of replication and standardization. Using standard equipment designs and transportable skid dimensions, modular designs can be made fit for purpose, scaled, or expanded as needed. Modularization can also have a significant impact on changeout times and the amount of sparing required. This presentation looks at reactor systems that lend themselves to modularization, including their benefits/capabilities as well as their limitations/disadvantages.

(3a) Use of Cogeneration in Large Industrial Projects

Conference: Southwest Process Technology Conference

Year: 2020

Proceeding: 12th Southwest Process Technology Conference

Group: Southwest Process Technology Conference

Session: Energy Management & Efficiency/TIEEP Energy Forum

Time: Thursday, October 1, 2020 - 10:30am-10:55am

Authors:

Loney, J., Fluor

This presentation looks at the extent of the use of cogeneration in recent large industrial projects executed by Fluor. For revamp projects, how are any increased requirements for steam and electric power addressed? For new facilities, to what extent is cogeneration being included in the design? What are the considerations when deciding whether to implement cogeneration? A brief overview is given describing the relative energy efficiency gains when using cogeneration compared to other strategies for supplying steam and electric power.

(3b) Optimum Design and Operation of Complex Steam Systems at Oil & Gas Industries through Combined Heat and Power Optimization Model

Conference: Southwest Process Technology Conference

Year: 2020

Proceeding: 12th Southwest Process Technology Conference

Group: Southwest Process Technology Conference

Session: Energy Management & Efficiency/TIEEP Energy Forum

Time: Thursday, October 1, 2020 - 11:00am-11:15am

Authors:

Owaidh, M.
Dulaijan, A., Saudi Aramco
Hazazi, A., Saudi Aramco

The steam and power systems in oil, gas, refining & petrochemical facilities are very complex and require optimization models to support operation to achieve required objective while meeting all constraints with the least operating cost. For the subject objective, Saudi Aramco has developed a combined heat and power (CHP) optimization model for each site. These models were found to be extremely useful in understanding the interactions between the various utility's major components. These components include boilers, cogeneration units, steam turbines, steam system network including process steam users and BFW condensate system. The interactions between these components create areas for energy optimization opportunities that lead to energy savings, GHG emission reduction and operating cost reduction. The mode considers key design and operational constraints such as: minimum steam reserve, equipment minimum and maximum operation limits as well as ensuring steam balance and energy balance is closed throughout the analysis. In addition, CHP optimization model for new design has been deployed at Saudi Aramco new facilities with energy savings benefits exceeding 10% from base case design. This paper covers several case studies addressing both operational and design improvements via utilizing the CHP optimization model, includes:

1. Design:

- Evaluating the economics of various Cogeneration options

- Evaluating the optimum sizing of a new STG

2. Operational:

- Evaluating the impact of Boilers load management
- Identifying the best combination between steam turbines and motors meeting process pumping and compression requirements with the least operating cost.

(3c) CHP, Energy Resilience, and Grid Reliability

Conference: Southwest Process Technology Conference

Year: 2020

Proceeding: 12th Southwest Process Technology Conference

Group: Southwest Process Technology Conference

Session: Energy Management & Efficiency/TIEEP Energy Forum

Time: Thursday, October 1, 2020 - 11:30am-12:00pm

Authors:

Dillingham, G., HARC
Gamarra, C., HARC

Combined Heat and Power (CHP) is a clean, flexible, robust on-site generation technology. With CHP an organization can improve on-site power resilience, thereby reducing business disruptions, while at the same time reduce overall power costs. Further, CHP does not only providing on-site benefits, but also can support the entire power grid through the provision of ancillary services, load shedding and shaping, and power export, as well as supporting grid-wide renewable energy integration. Many of these services are particularly important in locations with grid congestion and price volatility. Join the US Department of Energy's Southcentral CHP Technical Assistance Partnership team, Dr. Carlos Gamarra, P.E. and Dr. Gavin Dillingham, to discuss how CHP can significantly improve energy resilience and grid reliability through strategic, flexible operation of the CHP plant. Further, learn how CHP is becoming the go to resource to support renewable energy integration in both on-site microgrids and the power grid.

(4a) Tower Doctor(TM) Service from Koch-Glitsch: Solving Challenging Tower Problems

Conference: Southwest Process Technology Conference

Year: 2020

Proceeding: 12th Southwest Process Technology Conference

Group: Southwest Process Technology Conference

Session: Distillation & Separation

Time: Thursday, October 1, 2020 - 1:30pm-2:00pm

Authors:

Nieuwoudt, I., Koch-Glitsch
Rizo, M., Koch-Glitsch LP

The TOWER DOCTOR™ service from Koch-Glitsch provides technical assistance to solve complex problems in operating units in various industries. This presentation will showcase two examples were these services were utilized. The TOWER DOCTOR™ service team recently worked on a petrochemical tower in the US that was experiencing problems with capacity and efficiency. The team used CFD Studies, Gamma Scan and CAT Scan data, simulation support and on-site assistance to quickly identify the root cause of the problem. Equipment and installation services were provided to get the tower to where it is running very well. In another project, the team worked on a refinery tower in Europe that was having capacity and separations limitations. In this project, the team provided simulation support, gamma scan procedures and analysis, equipment and installation support. The tower now works well.

(4b) Fair Comparison of Packing Performance Data from Different Reference Systems

Conference: Southwest Process Technology Conference

Year: 2020

Proceeding: 12th Southwest Process Technology Conference

Group: Southwest Process Technology Conference

Session: Distillation & Separation

Time: Thursday, October 1, 2020 - 1:55pm-2:20pm

Authors:

Flowers, J., DowDuPont
Mukherjee, S., Dupont
Lyon, M., Dupont

Engineers are responsible for insuring their company makes wise decisions when purchasing distillation column internals. This task is often complicated by a lack of "apples to apples" performance data on competing products - in which there are a common reference system, similar operating conditions, and equivalent-quality peripherals.

The authors have attempted to use available performance data and published correlations for mass-transfer and capacity for random packing to create a fair comparison basis across different reference systems. This presentation will focus on the development of a simple transformation which gives useful insight into the likely capacity of a packing in any system, given data from another system.

We have also attempted, with less success, to develop a similar transformation for mass transfer comparisons. These charts will also be included, but available time will not permit discussing them in detail during this presentation.

(4c) Dividing Wall Column for Upgrading Naphtha Pool

Conference: Southwest Process Technology Conference

Year: 2020

Proceeding: 12th Southwest Process Technology Conference

Group: Southwest Process Technology Conference

Session: Distillation & Separation

Time: Thursday, October 1, 2020 - 2:30pm-2:55pm

Authors:

Bhargava, M., DWC Innovations LLC.

The refinery has an existing a naphtha splitter followed by a Depentanizer column. The light Naphtha was separated into to a mix of C5 cut and C6+ naphtha. The new business objective required to produce three products: iC5, C5s & C6+ naphtha.

The paper explores the common distillation sequences to meet the new design objectives:

- Install a new deisopentanizer downstream of the existing depentanizer
- Revamp the existing depentanizer to a conventional side cut column
- Revamp the existing depentanizer to a dividing wall column.

The pros and cons in terms of CAPEX and OPEX are discussed for the each of the revamp scenarios.

(5a) Good Practices for the Control of Hazardous Waste Emissions

Conference: Southwest Process Technology Conference

Year: 2020

Proceeding: 12th Southwest Process Technology Conference

Group: Southwest Process Technology Conference

Session: Process Safety

Time: Thursday, October 1, 2020 - 1:30pm-1:55pm

Authors:

Walters, M., Exponent, Inc.
Dee, S., Exponent
Cox, B. L., Exponent, Inc.

Proper management of waste streams generated by hazardous materials is important for ensuring worker safety and reducing environmental impact. This presentation explores good practices related to mitigating complex vapor mixtures emitted from hazardous wastes. Guidance is provided on developing and testing appropriate technology or technologies for the removal or destruction of unique hazardous vapor mixtures. Mitigation strategies for emissions from the Hanford Tank Farms, which contain millions of gallons of radioactive chemical waste in 177 underground storage tanks, will be used as a case study to demonstrate the do's and don'ts of implementing a novel technology for the control of hazardous waste emissions.

(5b) Mitigating Low-Temperature Metal Embrittlement during Overpressure Scenario

Conference: Southwest Process Technology Conference

Year: 2020

Proceeding: 12th Southwest Process Technology Conference

Group: Southwest Process Technology Conference

Session: Process Safety

Time: Thursday, October 1, 2020 - 1:55pm-2:20pm

Authors:

Arora, G., Bechtel Oil & Gas
Thakore, S., Bechtel Oil, Gas and Chemicals
Chittibabu, H., Bechtel Oil, Gas and Chemicals

High pressure natural gas flashed to atmospheric pressure result in low temperatures due to Joules Thomson effect. In LNG or gas plants there are systems and piping which are exposed to low temperatures during relief scenario.

Tube rupture is a commonly considered relief scenario for shell and tube exchangers. In closed loop refrigeration systems, a high pressure natural gas typically flows through the tube side and a low pressure refrigerant flows through the shell side. During a tube rupture scenario, the natural gas mixes with refrigerant and is relieved through a relief valve located on the shell side of the exchanger to prevent catastrophic equipment failure due to overpressure. However, the exchanger and piping must also be protected against low temperature embrittlement which can potentially be caused by natural gas as it gets cold while expanding from high pressure to low pressure.

Based on normal operating conditions the initial material selection was done based on the typical standard metallurgy used in the industry for this exchanger and piping. However, a steady state analysis of the tube rupture scenario revealed that the metallurgy may not be able to handle the colder temperatures generated during this scenario. This prompted several discussions relating to the impact of this finding such as:

- Are steady state results overly conservative?
- How much of the metallurgy will get impacted?

- How much is the time to consequence for this scenario?

In this presentation we will show how advanced tools such as computational fluid dynamics (CFD) and dynamic simulation were used to analyze the scenario and provide mitigations. The presentation will illustrate the temperature profiles, the critical areas, and where and how mitigation was implemented.

(5c) Need for Re-Evaluation of Relief System Design Under Current "New Normal" As Certain Credits Disappear

Conference: Southwest Process Technology Conference

Year: 2020

Proceeding: 12th Southwest Process Technology Conference

Group: Southwest Process Technology Conference

Session: Process Safety

Time: Thursday, October 1, 2020 - 2:15pm-2:45pm

Authors:

Saraykar, S., Ingenero Inc.

We have seen a number of economic factors collide recently: steep drops in demand and prices across basic commodities including oil and petrochemical products, uncertainty across global economies, and mixed actions from economic and political leaders of many countries. This can be attributed to the varying degree of spread of COVID-19 across the globe. All these conditions have placed companies under economic stress, especially those in the process industry struggling to maintain sales and output levels. Companies are taking radical actions to reduce capital and operational expenditure. These economic pressures can be taxing on all areas of a business, but especially on the production floor-level operations.

Cost-cutting directives are coming from the executive levels, demanding a significant reduction in expenditures while maintaining ongoing production. While all areas of a company feel the impacts, key process safety impacts are experienced on facility floors. A slight slip in the level of process safety endangers operators, maintenance personnel, and their supervisors. During these times of extreme economic pressure, it is imperative that process safety management remains in full-fledged effect. '˜Safety first' must be an overriding motto and an unflinching standard for process manufacturing industries.

Numerous operating companies globally are faced with major unparalleled challenges. Some of the steps taken to address these challenges can compromise process safety leading to hazardous conditions. One of the steps taken involves reduction in number of on-site personnel owing to the COVID-19 response plan. API Standard 521 allows for operator intervention credit to prevent some of the overpressure scenarios from occurring. This presentation discusses all such overpressure scenarios where operator intervention credit may have been taken in the current design but with limited resource availability, relying on operators to avoid such overpressure scenarios can cause an overload situation and compromise overpressure protection.

(7a) Energy Conservation in Distillation

Conference: Southwest Process Technology Conference

Year: 2020

Proceeding: 12th Southwest Process Technology Conference

Group: Southwest Process Technology Conference

Session: Chemical Process Technology

Time: Friday, October 2, 2020 - 10:30am-10:55am

Authors:

Bravo, J., Fractionation Research, Inc.
Jewell, D., The Dow Chemical Company
Pendergast, J., The Dow Chemical Company
Vickery, D., The Dow Chemical Company

Energy Conservation in Distillation

John Pendergast, Dow Chemical Company (Retired)

Dennis Jewell, Dow Chemical Company

David Vickery, Dow Chemical Company

Jose Bravo, Fractionation Research Inc.

Separations operations account for roughly 50 to 70% of the energy and the capital in the large scale Chemical Process Industries (CPI) [1]. Not surprisingly, along with the consumption of the energy, separation processes also account for the same fraction of the capital required for the processes. By far the dominant unit operation utilized for these separations is distillation, along with the related operations of absorption and stripping*. This fact was recently brought out again in the widely read and highly regarded publication *Nature* by David Scholl and Ron Lively. [2] Lively and Scholl describe important separations that, if the separation could be performed in a more energy manner, could have worldwide impact. In this article, we briefly discuss some of the reasons that progress in energy conservation seems sometimes to be slow, and then we present ideas that have proven useful in reducing energy consumption in existing facilities.

(7b) Changing Compositions Impacts Pressure Control

Conference: Southwest Process Technology Conference

Year: 2020

Proceeding: 12th Southwest Process Technology Conference

Group: Southwest Process Technology Conference

Session: Chemical Process Technology

Time: Friday, October 2, 2020 - 10:55am-11:20am

Authors:

Theunick, G., The Dow Chemical Company

Changing Compositions Impacts Pressure Control

Author: Greg Theunick, Dow Chemical

This paper describes a case study of a multi-pass distillation process with uncontrollable tower pressure.

A chlorosilane intermediate material is purified via distillation campaigns on a shared asset. Customer requirements for higher quality resulted in a need to process the material twice through the same towers to remove different impurities. During the second pass the tower pressure was uncontrollable and the plant has lived with this for years. Pressures would rapidly swing up and down causing dramatic shifts in column temperature profile and inventories inside the tower. This has resulted in low yield and product quality challenges. Several attempts were made to resolve the issue but with no success.

A thorough investigation revealed the composition of the feed during this second pass was not compatible with the installed pressure control scheme. The tower runs at vacuum pressure for this product and the control scheme requires the presence of a higher volatility species to hold tower pressure. The volatile components are removed on the first pass leaving very little to hold pressure steady during the second. Once this was identified, the addition of a noncondensible was included in the pressure control scheme and this instantly resolved the problem. The process has been able to achieve record yields and reduce significant costs. No capital was needed to resolve this long term issue.

(7c) Understanding Exchanger Practically for Improving Performance

Conference: Southwest Process Technology Conference

Year: 2020

Proceeding: 12th Southwest Process Technology Conference

Group: Southwest Process Technology Conference

Session: Chemical Process Technology

Time: Friday, October 2, 2020 - 11:30am-12:00pm

Authors:

Thorat, S., Ingenero Inc.

Exchangers are one of the key assets of almost any operating unit and an important aspect of overall process performance improvement. Optimization of exchanger operation is critical and is feasible through application of the right predictive and prescriptive analytical techniques.

A good monitoring process continuously tracks equipment closely to identify deviations in operating parameters and understand issues before they impact performance. Nevertheless with multiple process parameter changes it's not easy. Using advanced analytical tools in combination with machine learning techniques and AI is a valuable and effective way to identify deviations and problems earlier than traditional manual efforts. Taking a pro-active technical approach anticipates the problem before it reaches alarm levels and generally provides time for correction.

This paper will discuss techniques to identify deviation in equipment performance easily and accurately and provide insights quickly. Additionally, the paper will describe how predictive and prescriptive analytics can help determine the impact of deviations, the need for corrective action and assist in reaching innovative solutions.

Exchanger performance troubleshooting can be a tedious process as multiple operating and design factors affect actual performance. Sometimes smart design improvements are needed, but often options are available to restore the desired performance through intelligent process adjustments. Ensuring optimal exchanger operation requires understanding of the design/process complexities and overall functioning of the equipment.

This paper will cover several case studies to demonstrate on how units were able to substantially enhance their Overall Asset Effectiveness by improving the exchanger monitoring and performance.

(8a) U.S. CAPEX: A Post-COVID World

Conference: Southwest Process Technology Conference

Year: 2020

Proceeding: 12th Southwest Process Technology Conference

Group: Southwest Process Technology Conference

Session: Petroleum Refining Technology

Time: Friday, October 2, 2020 - 10:30am-10:55am

Authors:

Nichols, L., Hydrocarbon Processing / Gas Processing

Many aspects of the hydrocarbon processing industry (HPI) have changed since COVID-19. Despite significant declines in fuels and petrochemicals demand globally, future consumption is forecast to rebound in 2021 and beyond. The same trend is also apparent in global capital expenditures. This presentation will detail the current state of the global HPI projects construction industry, with a focus on capital spending in the U.S. midstream and downstream processing industries.

(8d) Integrated Refinery and Petrochemicals

Conference: Southwest Process Technology Conference

Year: 2020

Proceeding: 12th Southwest Process Technology Conference

Group: Southwest Process Technology Conference

Session: Petroleum Refining Technology

Time: Friday, October 2, 2020 - 10:55am-11:20am

Authors:

Turner, J., Fluor Corporation

Numerous studies have documented that integrated refinery and petrochemicals facilities, that process crude oil into fuels and petrochemicals, have a higher return on investment. This presentation takes a fundamental look at what products can be produced starting with crude oil, and what decisions need to made to develop the best facility configuration. Advantages and disadvantages of the options are discussed, and a potential overall configuration is developed.

(8c) Digital Transformation in Refining: Opportunities and Pitfalls

Conference: Southwest Process Technology Conference

Year: 2020

Proceeding: 12th Southwest Process Technology Conference

Group: Southwest Process Technology Conference

Session: Petroleum Refining Technology

Time: Friday, October 2, 2020 - 11:15am-11:45am

Authors:

White, D. C., Emerson Process Management

Almost every current software initiative in the refining and petrochemical industries seems to be called "digital transformation" with the implication that it is a completely new approach. However, the digital transformation of the refining and petrochemical industries has a long and rich history with many long-term successes and some examples that were less than fully successful. In this presentation there will be a short review of the history of these developments, the trends that have characterized its evolution and the lessons learned that can help us with evaluation of the many new technology options available today. There are many claims with regard to the potential benefits of these new technologies such as big data analytics, AI and IIOT, some with totally unrealistic values. They all require investment and manpower and their likely payout needs to be analyzed rigorously versus other investment alternatives in the context of actual operating goals. Significant additional value and margin can be gained by improving the accuracy of forecasts of future plant behavior, including potential production and supply chain alternatives, and early detection of potential availability problems and product quality issues. Improvements in safety and environmental performance can also be obtained. In this presentation a structure will be presented for characterization and evaluation of the technologies and their financial analysis. Actual case studies will be used to illustrate the impact of some of these new tools on plant productivity and margins.

(9a) Image Classification to Automate at-Line Pellet Inspection in Plastics Production

Conference: Southwest Process Technology Conference

Year: 2020

Proceeding: 12th Southwest Process Technology Conference

Group: Southwest Process Technology Conference

Session: IIoT/Big Data

Time: Friday, October 2, 2020 - 1:30pm-1:55pm

Authors:

Peng, Y., Dow Inc.
Braun, B., The Dow Chemical Company
Broadway, M.
McAlpin, C., The Dow Chemical Company
Colegrove, B., The Dow Chemical Company
Chiang, L. H., The Dow Chemical Company

For many applications it is critical to deliver contamination free polyethylene product. Therefore, continuous inspection of polymer production is vital. Most systems inspect a pellet slip stream in an at-line fashion and provide feature information about any present contaminant such as color and size. However, in certain scenarios it is also of interest to determine whether the contamination is free flowing in the pellet stream (loose) or incorporated into the polymer pellet (embedded). Typical analytical equipment does not provide this information and the classification is thus a manual and subjective task.

In order to automate this classification, a multi-class classifier was built with a convolutional neural network (CNN). The network layers will automatically learn complicated features from raw images compared to traditional machine learning methods, which require well-defined features based on domain knowledge about the output class. A selection of well-established CNN model structures (i.e. VGG16, ResNet50, etc.) were tested with and without image augmentation and model accuracy evaluated. The best performing model achieves greater than 95% accuracy for all output classes. This is a successful application of deep learning techniques to directly improve manufacturing efficiency and the model is in use for daily decision making.

(9b) A Roadmap to Achieve Operational Excellence through Advanced Data Analytics

Conference: Southwest Process Technology Conference

Year: 2020

Proceeding: 12th Southwest Process Technology Conference

Group: Southwest Process Technology Conference

Session: IIoT/Big Data

Time: Friday, October 2, 2020 - 2:00pm-2:15pm

Authors:

Majumdar, A., Ingenero, Inc.
Thorat, S., Ingenero Inc.

Asset Optimization and Operational performance can be improved significantly by using advanced analytical tools such as Big Data Analytics, IIoT/P, and AI in combination with fundamental models by developing a "**Digital Twin**".

Ingenero uses the "**Digital Twin**" approach to identify operational gaps and effective resolution of issues in timely manner to maximize performance. Ingenero's Excellence Through Insight program is successful through combining the right information with in-depth domain expertise and deep data analytics to generate predictive and prescriptive recommendations presented in customized visual displays and dashboards for intelligent decision making.

The success of the process requires the appropriate volume of data. Post collection and mining of data, tools are used for effective reconciliation, validation and consolidation of data to gain insights quickly. This information is utilized as an input for machine learning models and these models convert the raw information into useful statistics/analytics that maximize asset effectiveness.

This paper will discuss the Digitalization process utilizing fundamental models combined with data analytics techniques to achieve operational excellence. The appropriate techniques, tools and equipment models that are applied to track the performance and identify the size of deviation and improvement opportunities on real-time basis will be discussed. Additionally, the paper will describe how predictive and prescriptive analytics can help determine the impact of deviation, need for corrective action and assist in reaching challenging solutions.

The Industry 4.0 Digitalization initiative is driving industry specific advanced analytics to leverage both historic and real-time data to achieve improvements that are difficult to accomplish using traditional approaches. Substantial economic gains can be achieved when data analytics is coupled with the overall plant operating strategy.

This paper will cover some of case studies, to demonstrate on how Manufacturers were able to improve their Overall Plant Performance and Capacity by 5 - 15%.

(9c) BHC3 Reliability for Process Manufacturing

Conference: Southwest Process Technology Conference

Year: 2020

Proceeding: 12th Southwest Process Technology Conference

Group: Southwest Process Technology Conference

Session: IIoT/Big Data

Time: Friday, October 2, 2020 - 2:30pm-2:55pm

Authors:

Holtan, T., C3.ai

Reliability of processes and equipment, such as Steam Crackers, in downstream manufacturing depends on critical and noncritical equipment avoiding both expected and unexpected failure modes. The goal is to enable to Operators to use actionable insights from advanced machine learning models to gain early warning of mechanical and process failure modes. Early warning decreases expenses by shifting unplanned maintenance to planned and increases revenue by improving process availability.

(10a) Production of Renewable Fuels: R100 and Co-Processing

Conference: Southwest Process Technology Conference

Year: 2020

Proceeding: 12th Southwest Process Technology Conference

Group: Southwest Process Technology Conference

Session: Sustainable Technology

Time: Friday, October 2, 2020 - 1:30pm-1:55pm

Authors:

Rasmussen, H., Haldor Topsoe
Gabrielsen, J., Haldor Topsoe A/S

The presentation will discuss the current trend in renewable fuels regulation in the US and the importance of the Carbon intensity score (CI) for the renewable fuels. We will review various types of renewable feedstocks and the impact to the design of the technology with respect to material selection and requirement for pretreatment. The presentation will also discuss the reaction pathway and chemistry for production of renewable diesel and the required dewaxing to meet the cold flow properties. We will show two case stories of co-processing of renewable fuels and fossil feed in the same unit and discuss the required catalyst selection.

(10b) An Industrial Gas Supplier's View in an Evolving Low Carbon Market and Low CI Score for Renewable Diesel

Conference: Southwest Process Technology Conference

Year: 2020

Proceeding: 12th Southwest Process Technology Conference

Group: Southwest Process Technology Conference

Session: Sustainable Technology

Time: Friday, October 2, 2020 - 1:55pm-2:20pm

Authors:

Holland, J., Air Liquide
Phillips, D., Air Liquide

Optimizing renewable diesel production for maximum profits requires the dual objective of lowering carbon intensity and the utilization of co-products for high value applications.

The presentation will provide an overview of Air Liquide Engineering and Construction technologies for refinery efficiency focusing on renewable diesel feedstock pretreatment, hydrolysis, carbon capture, and production of renewable chemicals. In addition, an overview of the Air Liquide Large Industries hydrogen pipeline and its low carbon attributes will be explained. Finally, the added value to the refinery from freeing up co-products will be explained with the objective of enabling producers to maximize their profits and lower carbon footprints.

(10c) Pyrolysis: The Renewable Plastics Solution

Conference: Southwest Process Technology Conference

Year: 2020

Proceeding: 12th Southwest Process Technology Conference

Group: Southwest Process Technology Conference

Session: Sustainable Technology

Time: Friday, October 2, 2020 - 2:15pm-2:45pm

Authors:

Crawford, S., Agilyx

The world's plastic production is growing at 9% annually, yet only around 10% of these plastics ever get recycled. Plastics, especially post-consumer, can be challenging to recycle due to their variability and contamination, product requirements, and supply chain complexity. As an essential part of our daily lives, it is critical to improve upon the renewability of the plastics lifecycle.

Agilyx has integrated five pillars to address these recovery and recyclability challenges to ensure a renewable plastics solution. Through a proprietary technology platform using integrated feedstock management, research and development, data acquisition and analytics, and predictive modeling, Agilyx can formulate solutions to generate targeted products within base feedstocks. These include virgin-like polymers, low-carbon fuels, and chemical base stocks and intermediaries. The characterization of the chemistry within post-use plastics supports multitudes of pathways for products.

Creating renewable plastics through pyrolysis opens the doors to this largely unutilized resource to help mitigate the need to rely solely on conventional fuels and chemical feedstock production. Circularity for renewable plastics promotes environmental stewardship to manage a lower carbon intensity and greenhouse gas emission. When compared to conventional chemical production, this results in the reduction of the use of virgin materials, promotes the reuse of polymers, and creates an indefinitely recyclable loop.

AUTHOR INDEX